Anonymous

Intravenous transfusion

Of saline solution with a new apparatus

Anonymous

Intravenous transfusion
Of saline solution with a new apparatus

ISBN/EAN: 9783337900137

Printed in Europe, USA, Canada, Australia, Japan

Cover: Foto ©berggeist007 / pixelio.de

More available books at **www.hansebooks.com**

Intravenous Transfusion of Saline Solution with a New Apparatus.

BY

ELY VAN DE WARKER, M. D.,

SYRACUSE, N. Y.

Reprinted from the New York Medical Journal
for December 12, 1896.

INTRAVENOUS
TRANSFUSION OF SALINE SOLUTION
WITH A NEW APPARATUS.

By ELY VAN DE WARKER, M. D.,
SYRACUSE, N. Y.

THE intravenous transfusion of saline solutions is such a simple operation that ordinarily no special apparatus is necessary. Every surgeon knows, however, that it is sometimes these very simple procedures that are difficult. In a case of extreme collapse, when the veins have no existence, physiologically speaking, and exist only anatomically as collapsed tubes, without radial pulse—when no amount of cording will cause the empty vein to fill, and when a lost minute may be measured by a lost life—it is very convenient to have a simple apparatus at hand that will expedite even so simple an operation. In these extreme cases I believe it is necessary to transfuse into the veins; subcutaneous transfusion is too slow with the vital forces so nearly suspended. Saline rectal enemata, a method of great value, for the same reason is too slow to furnish the necessary stimulus to a heart nearly inert for the need of a fluid to act upon. The apparatus here figured is a composite affair that was gradually put together from time to time as the need of transfusion occurred, until it reached its present perfected form.

While nearly every surgeon has had occasion to test the merits of saline transfusion, I doubt if any one has had the satisfaction of demonstrating its value as a lifesaver more thoroughly than I have had. With the object of inducing some one who has never tried it to be prepared to use it with a strong reliance upon its certainty of action, I will narrate briefly the following cases:

CASE I.—Miss M., of Irish nativity, aged forty-three years, large ovarian cyst of several years' duration, had suffered no unpleasant symptoms other than those due to abdominal distention. While nursing a sick relative, which caused very unusual exertion, she was suddenly attacked with severe abdominal pain and tenderness, with chills and subsequent fever. I saw her at the home of her relative about one week after the outbreak. The abdomen was of a dimension equal to a pregnancy at term, very tender on palpation, with frequent attacks of pain. Temperature, 103°; pulse, 110. I advised immediate operation, but it was nearly a month after my visit that she was admitted to the Central New York Hospital for Women for operation. During that interval the symptoms had gradually augmented, and at the time of her admission she was in a deplorable condition. She was greatly emaciated and too weak to walk. Section showed a monocyst almost universally adherent. The contents of the cyst were about twenty pounds of extremely fœtid pus. The adhesions were easily broken down, and the operation in particular difficult, lasting about forty minutes to the completion of the toilet. She was taken off the table in fair condition. It was soon noticed by the nurse that she was growing cold and the pulse raeing, and in a few minutes she appeared to be in a dying condition—pulseless, gasping respiration of the dying. with complete surfaee anæsthesia. Intravenous transfusion of a one-per-cent. salt solution at 105° was at onec made, using about twenty ounces of fluid. In this case it was very difficult to find and enter a vein, so complete was the collapse of the vessel. My troear and cannula greatly expedited the operation. The result of transfusion was prompt and lasting and the patient's life was saved. The operation was very nearly bloodless, the adhesions showing no disposition to bleed as they were broken up. The eonditions leading to collapse in this ease were interesting, as demonstrating that it is not blood loss previous to, during, or subsequent to operation, but that any condition equivalent to blood loss will preeipitate the moment of eollapse. Here high temperature, arrested nutrition, and blood poisoning were those equivalent conditions.

CASE II.—A woman, aged thirty years, married, demiblonde. number of children unknown. In the service of one of the staff of the Woman's and Children's Hospital of Syracuse. Abdominal seetion reaehed a right pelvie hæmatoma due to what was supposed to be an extrauterine pregnancy. An old elot was turned out of the pelvis equivalent to about a pound in weight. It was dense and tarlike in color, and at the time the lesion oecurred. must have represented a quart, or possibly more, of blood. The phlegmon developed about six weeks previous to the operation, but its history was obseure. The loss of blood due to the operation was small. The patient was upon the table about forty minutes. After being put to bed in a fair eondition of pulse, she surprised us by passing quiekly into extreme collapse. She was algid, with complete surfaee anæsthesia, eye and other reflexes abolished, pulseless, and the veins eompletely eollapsed. The respiration was gasping and shallow. At the request of the operator I performed transfusion. I was about to abandon the operation on aeeount of the impending death, but as she was yet gasping and the trocar nearly in the vein, I persevered, and in less than a minute the solution was being transfused. The effect was rapid, and the patient passed from a condition of inevitable death to one of comparative safety in about three hours. She made an uneventful recovery.

CASE III.—As anæmic shock is not the only form with which the surgeon has to contend, it is well to refer briefly to the following case with a view to contrast:

Miss ——, aged twenty-one years, single, had always been a delicate girl. At seventeen years she had not yet menstruated. Her physician detected a pelvic mass extending like a ridge from side to side. As she was bedfast and rapidly failing, laparotomy was advised and consented to. She was admitted into the Woman's and Children's Hospital of Syracuse. The mass was made up of the lower border of the omentum fusing into the right tube and ovary, all of which were removed. She made a prompt recovery, and improved very considerably in general health. The pelvic condition was regarded as tubercular, although I am not sure that it was verified by the microscope. Four years after she was readmitted with symptoms of intestinal obstruction. At the request of her physician, a colleague on the staff, I did the operation. The intestines, large and small, were greatly distended. At first it was supposed that the obstruction was at the sigmoid, but in the Trendelenburg position it was seen that what appeared to be a constricted part of the colon was a cyst of tubal origin which was tied off. At this point a serious accident occurred. The intestines, which had been retained in the abdominal cavity only by great care and watchfulness, through the momentary inadvertence of the assistant were allowed to escape. As usual in such a case, the intestines were reduced with extreme difficulty and loss of time. They were at last returned and the search for the obstruction resumed. It was found not far from the ileo-cæcal region, and consisted of a globular mass surrounding and obliterating the intestine. It was found necessary to resect the intestine at a healthy portion and insert a Murphy button. The operation lasted about an hour. She was put to bed in fair condition; pulse about 120; respiration, 18; temperature, 99.5°. This was at 3 P. M. While at dinner in the evening I was called to the hospital by my colleague in consultation in the case. I saw her at 7 P. M. Temperature, 104°; respiration, 30; pulse racing. At 8.30 P. M., pulseless at the wrist; temperature, 104.6°; respiration, Cheyne-Stokes. At the request of my colleague I made transfusion in the hope that an increased volume of fluid would stimulate the fast failing heart. Owing to the carelessness of a nurse the amount transfused is not known. It had no effect upon the circulation. She died at 10 P. M. with a temperature of 106°.

The case is given somewhat in detail as it demonstrates in a striking manner the limits of saline-solution transfusion. The three cases serve to define three forms of shock. First, we find shock due to inanition, fever, and chronic sepsis, precipitated by operation. Second, shock due to blood loss prior to operation, and both forms to amend to transfusion. In the third, the fatal condition, or shock, comes from suspension of inhibitory nerve centres that functionally control the heart and heat centres. The train of conditions that led up to this is clear when we recall the eventration, and all that it implied to the nervous system. The entire nervous congregation of organic life was roused into fatal overstimulation through the irritation of the greater and lesser splanchnic nerves, and the brain and spinal centres implicated through the great sympathetic, quickly resulting in suspension of inhibitory centres. Thus, while saline transfusion was able to supply the deficiency in loss of circulating fluid, it was unable to supply what was necessary in the other. Clinically the two types stand in marked contrast: In the first, advancing algidity; in the second, augmenting pyrexia. In the first, surface anæsthesia increases as surface heat is lost; in the second, only as the brain ceases to functionate. In the first, consciousness is lost in the same ratio as the pulse is lost and the heart's action fails; in the second, rather with the increasing pyrexia. In the first, torpor from the beginning, intensifying cold, death. In the second, restlessness, intensifying heat, death. One is as much deserving the name of shock as the other, and we have no other name for either. As intravenous transfusion has now a status as a life-saving operation beyond all controversy, it is well to be able to define just in which type of shock we may expect these results, and as a contribution to this knowledge I have ventured to overstep the limits of the paper and introduce the third case.

In several instances in which the apparatus has been exhibited before societies, several gentlemen suggested that just as prompt results could be gained from rectal or subcutaneous injection. I have no doubt that is so in cases in which function is not too seriously depressed. But in cases in which death is near and time a serious matter, I would not advise any to trust to it. I have had some experience that has practically taught me this lesson. I have found rectal hot saline injections of great value in supplementing intravenous transfusion when the radial pulse is re-established and the reaction appears to drag; then the injection appears to augment the effect of the transfusion. When reaction is established to this degree, the rectum in my experience offers

as favorable a channel as the subcutaneous route, and
leaves no painful sequelæ.

The physiological study of intravenous injections
has begun to awaken keen interest. MM. Bosc and
Vedel, in the *Journal des praticiens,* July, 1896, have
reached the following conclusions from their study of
the subject: Large intravenous transfusions are not
toxic even when transfused as rapidly as forty to eighty
cubic centimetres a minute. These injections produce
an abundant diuresis in half an hour after the transfusion. While this·is, no doubt, true in physiological
experiments, I have not observed it in cases of extreme
collapse in the human subject. There is no elevation
of blood pressure and no albuminuria, but there is increased heart action and an elevation of the central and
peripheral temperature—febrile reaction. They quote
Mayet that in the compound saline solution—chloride and sulphate of sodium—the sulphate is of no
value and should be avoided, as it affects the blood
globules, although the authors say there is no difference
between the effects of the two solutions. Fasting animals appear to be more susceptible to large injections,
but even when apparent death results the animals quickly recover. In these experiments it was shown that
the physiological effects were not in proportion to the
temperature of the solution and the rapidity with which
the injection was made. The authors recommend a
simple saline solution, one hundred and five grains of
sodium chloride to thirty ounces of water to be the pref-
erable proportion. In the hasty approximation of the
operating room, a teaspoonful to a quart of water is a
good working formula.

In the employment of the intravenous injections

after the hæmorrhage of typhoid fever—and it ought always to be used when collapse threatens—it has recently been shown that the beneficial effect is not confined to merely increasing the volume of the circulating blood. There is a far-reaching and profound effect that can not be measured in this mechanical way. M. Claisse, in the *Comptes rendus de la Société de biologie*, 1896, has made his study upon the blood from the fact that in many infectious diseases increased leucocytosis is a constant attendant. "In a case generalized purulent streptococcus infection an intravenous injection of fifteen hundred grammes of saline solution was followed in an hour and a half by a rise of temperature from 102.9° to 105.8°, while the number of red blood-globules to the cubic millimetre was reduced from 3,968,000 to 3,596,000, and the number of white blood-corpuscles from 13,547 to 7,804. In the course of three hours the temperature had declined to 98.6°. In the case of an old man of sixty-four years with a diffuse phlegmon of the arm, a subcutaneous injection of a litre of saline solution was followed by a diminution in the number of red blood-corpuscles from 3,565,000 to 3,255,000, and of the colorless corpuscles from 26,660 to 11,346. In a case of profound puerperal infection an intravenous injection was immediately followed by an alteration in the relation between the red and colorless blood-corpuscles—1 to 228 to 1 to 344." In the comment of the *Medical Record*, from which this extract is taken, it is pointed out that the depression of temperature, diminished leucocytes, and amendment of the symptoms of infection with a status of reaction are similar to those following the injection of antitoxine in diphtheria. "A splendid field for research is thus opened up, and one that possibly will render the animal serums useless in the treatment of specific infections." I am strongly of the opinion that in view of the enormous renal stimulation following intravenous injections of saline solutions, and the marked kidney implication in diphtheria, together with the profound effect of the injection upon the corpuscular elements of the blood, this will prove an accepted treatment of diphtheria in the near future. In typhoid fever, in view of this report by Claisse, the saline transfusion has a double indication in hæmorrhagic cases. In puerperal infection I shall certainly try it. In cases of the nearly functionless kidneys of eclampsia and with the poison-laden blood it is positively indicated, and, I believe, has already been tried, but I have lost the reference.

From this brief abstract of what has already been accomplished in the therapeutics of saline solutions transfused directly or indirectly into the blood, it is evident that its usefulness in the future may be broadened far beyond the expectations of those who have heretofore limited its use to the treatment of shock or hæmorrhage. In fact, to the physician, rather than to the surgeon, will we look for this more important demonstration of its utility. Simple as the operation of intravenous transfusion may be, yet an apparatus that renders it easy to perform, aseptic and accurate in operation, will contribute an important element to this more extended employment of the method. To the physician who is but seldom called upon to make any surgical operation, the little apparatus here figured will prove of great usefulness.

The apparatus consists, first, of a glass container large enough to hold from three pints to two quarts of solution (Fig. 1), fitted with a sufficient length of pure gum tubing given off from the bottom of the container. In the figure the tubing is represented as connected with the stopcock C, Fig. 2, with the cannula D, Fig. 2 attached, and in the condition in which it is when in use, the cannula inserted in the vein and the

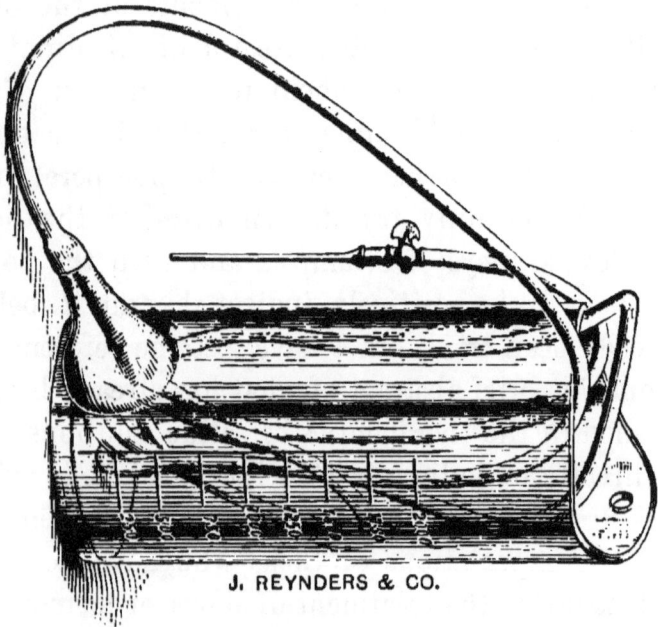

J. REYNDERS & CO.

Fig. 1.

stopcock open. The glass has a scale of ounces etched into its side, so that the quantity of fluid transfused at any stage of the operation may be readily ascertained. The second part of the apparatus (Fig. 2) consists of a cannula D, the trocar B, and the stopcock C. One end of the cock is corrugated so as to be firmly held by the distal end of the tubing. The other end of the cock, as seen in the figure, is slightly tapered so as to slip into the head of the cannula, which is beveled to correspond, so that when connected with the cannula an airtight taper joint is formed. The trocar B has a head sufficiently large to be easily grasped by the

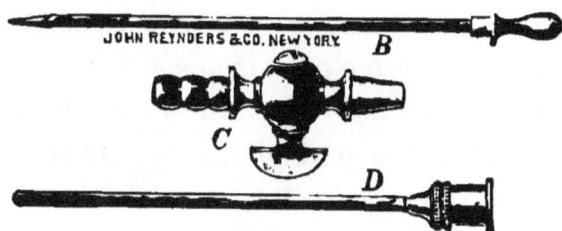

FIG. 2.

fingers and withdrawn when the cannula is inserted into the vein.

The operation is done by the use of this apparatus as follows: After the vein is exposed, the only instrument not here figured is a small mouse-tooth forceps to pick up the vessel, which is quite necessary; but any small forceps will serve the purpose. The container has been filled and the stopcock connected with the tubing is held by an assistant. The cannula and trocar together are inserted into the vein, which is held up by the forceps, so as to offer a slight shoulder to the sharp point of the trocar which is thrust into the lumen of the vein, being careful not to transfix. The stopcock is now opened, and the solution allowed to escape until the tubing is thoroughly warmed and the solution discharges at a proper temperature. You are now ready to withdraw the trocar and thus open the cannula, through which the blood flows more or less freely. The stopcock, still open and the solution discharging, is at once connected with the cannula by the taper joint— simply thrusting the two together. It is thus seen that the two currents, one of blood from the cannula and the other of saline solution through the cock, are joined without the possibility of the entrance of air. The flow of solution is regulated by the cock as well as by the height at which the container is held above the point of delivery. In the case of collapsed veins after a very exhausting hæmorrhage or extreme shock, no blood will escape from the cannula after the trocar is withdrawn. Fig. 2 represents that part of the instrument somewhat shorter than the actual length and about half the diameter.